PLUMBING REPAIRS SIMPLIFIED

by Donald R. Brann

Library of Congress Card No. 67–27691

FIFTH PRINTING
REVISED EDITION–1972

Published by
DIRECTIONS SIMPLIFIED, INC.

Division of
EASI-BILD PATTERN CO., INC.
Briarcliff Manor, N.Y.   10510

NOTE

Due to the variance in quality and availability of many materials and products, always follow directions a manufacturer and/or retailer offers. Unless products are used exactly as the manufacturer specifies, its warranty can be voided. While the author mentions certain products by trade name, no endorsement or end use guarantee is implied. In every case the author suggests end uses as specified by the manufacturer prior to publication.

Since manufacturers frequently change ingredients or formula and/or introduce new and improved products, or fail to distribute in certain areas, trade names are mentioned to help the reader zero in on products of comparable quality and end use.

The Publisher.

BE AN
INSTANT HERO!

A faulty plumbing fixture, a leaky faucet, and a deep seated headache have much in common — your peace of mind.

This book, like aspirin, can relieve many headaches. But unlike aspirin that offers relief, this book does much, much more — it provides cures, repairs that save money.

Those who follow the simplified directions and fix a leaky faucet, or "cure" a running toilet, soon discover the effort works wonders — on inanimate fixtures, as well as on individuals.

Not having to pay outrageous costs for simple repairs, learning how to do something they didn't believe themselves capable of accomplishing, the man of the house takes on glamour, becomes and Instant Hero.

Solving home repairs is an important part of family life. The man who takes care of his home, takes care of his family. And the home reciprocates. While "things" can't say "thanks," they do exert great influence on the lives of those within their sphere of activity.

Don R. Brann

TABLE OF CONTENTS

PLUMBING FACTS OF LIFE.

In many cities building codes only permit licensed plumbers to make specific installations and repairs. Those who want to learn how a repair is done, can insure getting and paying for work actually required, by reading this book. Knowing How—provides insurance against costly plumbing bills.

Read slowly. Note each illustration. If you are unfamiliar with plumbing, the illustrations and procedure will naturally seem strange. Don't let this bother you. On some mornings even your face can look strange.

Plumbing isn't nearly as difficult as most people imagine. All it requires is taking something apart—reversing the original process of assembly. If you know where part 1 is located, when, and how to remove part 2, 3, etc., then reassemble parts exactly as each illustration shows, you've got it made.

Every plumbing fixture in your home contains parts that were assembled in position shown in each illustration. To repair, or replace, reverse the process of assembly. Play safe. Number each part, or place small parts on a large calendar, or other numbered sheet. Place the first screw, nut or bolt you remove on 1, Illus. 1.

Through the years, plumbing fixtures have gone through many stages of design. Parts made for one faucet or fixture, seldom fit another. The key to its repair is simple. Just replace each part with its exact duplicate. Learning what part is required, is more than half the task of repair.

Your first step is to ascertain the manufacturers name and model number. While this takes a bit of sleuthing, you frequently find some identification, name or number, on the bottom of the fixture.

Be sure to get the right name. If it says American, is it just American, or American Brass, American Kitchen, American Standard, etc., etc.? If there is no identification, plan ahead.

If you don't know exact model number or approximate year fixture was installed, take a part, or snapshot of a fixture with you, and you have a good chance of getting what you need.

Keep tools required in a convenient place. Set up a vise where it can be used when needed. Remember, one plumbing repair bill can pay for a lot of tools. Two or more repairs, and you make sizeable savings.

Plumbing, like most crafts and trades, depends on sequence. What you take apart goes back together in the exact same position. While this may seem easy, a flat washer with a beveled edge may be installed with the face down or up. To be sure, note position of each part and place it on the numbered sheet with the proper face up.

WASHERS

Washers that "almost fit," seldom work. Get the exact size. Also the exact kind—rubber, fiber, nylon, leather, or the kind that was removed. Invest time getting acquainted with the fixtures described in this book.

Faucet washer sizes have been standardized. Illus. 2 shows

size and number. This simplifies ordering exact size. Buy hot
and cold washers that fit. Buy a few replacement screws.

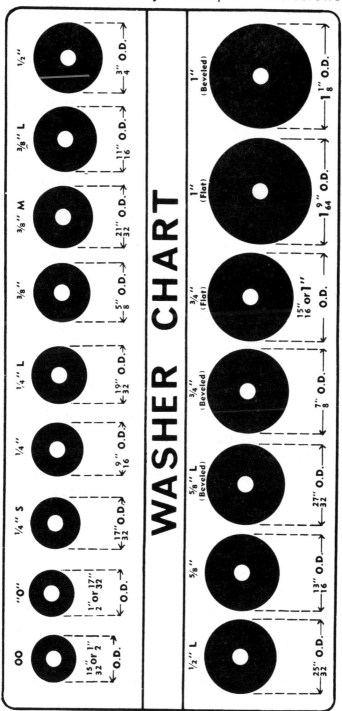

WASHER CHART

To play safe, remove the screw holding washer, Illus. 3, to see if it is still removeable. If not, a few drops of penetrating oil, (Liquid Wrench or equal) should loosen it. Or purchase a replacement spindle, Illus. 4.

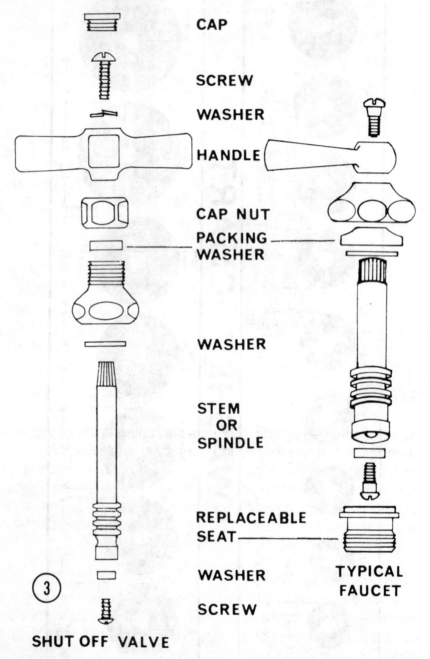

CAP

SCREW

WASHER

HANDLE

CAP NUT

PACKING WASHER

WASHER

STEM
OR
SPINDLE

REPLACEABLE SEAT

WASHER

SCREW

③ SHUT OFF VALVE

TYPICAL FAUCET

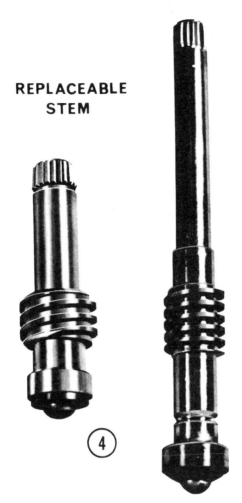

REPLACEABLE STEM

④

STEMS

Manufacturers assemble products according to their specialized methods, not according to industry standards. For this reason, step-by-step directions describe and illustrate basic repairs. Use it as a guide, and you'll soon know a ballcock from a toilet seal. Equally important, even if you make no repairs, you'll be able to talk intelligently to a plumber. If he knows you know what needs to be done, you effect substantial savings.

Learn the correct name of each replacement part. Talk like a "pro" when you walk into a plumbing supply house and they will accept you as a member of the inner circle. While busy clerks don't mind getting rid of a customer who doesn't know what he needs, they hesitate to louse up a "pro."

PLUMBING REPAIR FACTS

The actual cost of repair parts is small. It's the time required to go somewhere, find out what is needed, then go somewhere else to find and purchase the part required. If you don't tell the clerk exactly what you want, he'll guess to get rid of you as quickly as possible. This guess may be right, it can also be wrong. You'll get a part, you'll put it in, but it won't work.

Since other homeowners with the same problems are competing for the sales clerk's time and advice, your 10¢ washer or $2.00 gadget, can only take so many minutes to sell or retailer loses money on the transaction.

If you live any distance from a well stocked plumbing supply house, or it has more customers than its staff can conveniently serve, write the manufacturer. Tell him what equipment you own. Order a kit of repair parts.

Never attempt a plumbing job when you are tired, angry, or lack the time. If you "fear" doing it—that's OK. Always fear doing something you have never done before and you exert that extra something that keeps you alert.

Step-by-step directions in this book describe in detail how to:

1. Repair faucets.
2. Stop toilets from running.
3. Open up clogged toilets, sink and lavatory drains.
4. Repair broken or cracked toilet bowls and tanks.
5. Stop leaks in cracked pipes—and many, many other important plumbing repair jobs.

The plumbing in your home represents an important part of your total investment. Builders allocate 10% or more of the overall construction costs for plumbing. It can cost you double to replace plumbing after the house is completed.

As head of a household, it is your responsibility to establish operational rules for all equipment, and see to it they are followed. A waste paper basket should be first on your list for bathroom maintenance. Or if space permits, a small covered waste receptacle. Insist that all paper towels, bandaids, bandages, sanitary napkins, cigarette wrappers, cigar butts, etc., be deposited in the garbage can, not down the toilet.

Some teenagers vent jealousy and temper by deliberately throwing something down a toilet. Plumbers report the highest number of house calls in homes where teenagers entertain.

Learn where waste and water lines are located. Where to find cleanout plugs. This information can save you time, temper and money.

Don't allow anyone to drive nails in a wall that contains pipes. Copper tubing is especially easy to puncture. Since water will invariably follow a framing member, the actual leak frequently shows up some distance from its source.

If you find a leak where there's no fixture or pipe nearby, first ascertain whether anyone remembers driving any nails into a wall, and where it was done. In many cases, nails driven weeks or months previously can provide an important clue, even when the actual leak occurs far away.

The wall framing, Illus. 5, behind most kitchen and plumbing fixtures, contains 2x6 studs, placed 16" on centers. Studs may be notched to receive tubing, or holes drilled as shown, Illus. 5. Cats are usually nailed in position shown to support frame.

A nail frequently acts as a plug when driven into copper tubing. A leak can start months later, when the nail begins to rust. When you remove the nail, it accentuates the leak. If you find a nail driven into a wall containing copper tubing, use a screw driver to scrape out a small hole in plaster. If your sleuthing is accurate, you'll get water in your face.

Fix hole in pipe as specified on page 28, Patch hole with patching plaster. Paint wall and always hang pictures using paste-on hangers.

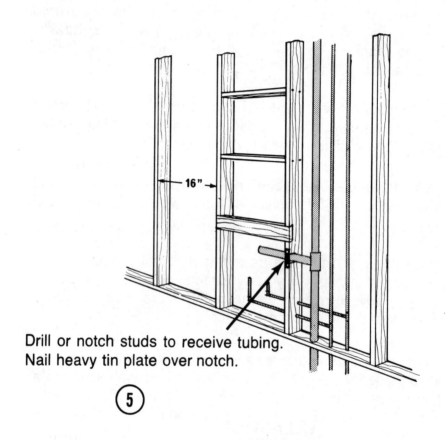

16"

Drill or notch studs to receive tubing. Nail heavy tin plate over notch.

(5)

SEPTIC TANKS

Those who buy homes in a sub-division, or in any area served by septic tanks should obtain a drawing, showing A, length and location of line between house and tank; B location of tank; C direction and length of drain fields, Illus. 6.

Also position of intake pipe in septic tank, Illus. 7. This information can save considerable money and labor when it's necessary to service tank.

14

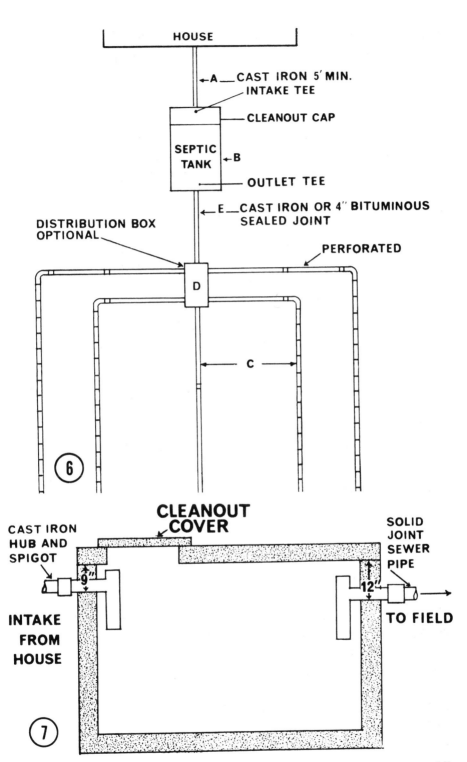

HOUSE

← A ___ CAST IRON 5' MIN.
INTAKE TEE

CLEANOUT CAP

SEPTIC
TANK ← B

OUTLET TEE

← E ___ CAST IRON OR 4" BITUMINOUS
SEALED JOINT

DISTRIBUTION BOX
OPTIONAL

PERFORATED

D

C

⑥

CLEANOUT
COVER

CAST IRON
HUB AND
SPIGOT

SOLID
JOINT
SEWER
PIPE

9"

12"

INTAKE
FROM
HOUSE

TO FIELD

⑦

15

Ask the builder or real estate agent to show you where clean out plugs, Illus. 8 are located in the main waste line, and in any subsidiary waste line from a sink or lavatory.

CLEAN OUT PLUG◄

PLUMBING TOOLS EVERY HOME SHOULD OWN.

Since the right tools simplify repairs, having tools available when needed, makes sense. While this book explains repairs anyone can accomplish by following step-by-step directions outlined, it's important to use the tools suggested.

Different size screwdrivers should be first on your list. A screw driver must fit and fill the slot in screw head. Unless it fits properly it tears the slot. Buy one screwdriver that fits the slot in a screw used to fasten washer to spindle, Illus. 3. Also buy screwdrivers for Phillips head screws, Illus. 9.

PHILLIPS
SCREW

Buy a can of penetrating oil, Liquid Wrench or equal. This now comes in spray, as well as in spout cans. The spray can is handy in hard to reach areas. Penetrating oil saves fixtures, parts, time and temper. Don't force anything. Use a little oil. Give it time to penetrate then tap the part gently to break a rust bond. Next use the tool required.

A stillson wrench Illus. 10, is handy in many places. It's especially handy if you don't happen to have a small vise. While a stillson or a pipe wrench isn't used often, it is one tool needed to hold pipe.

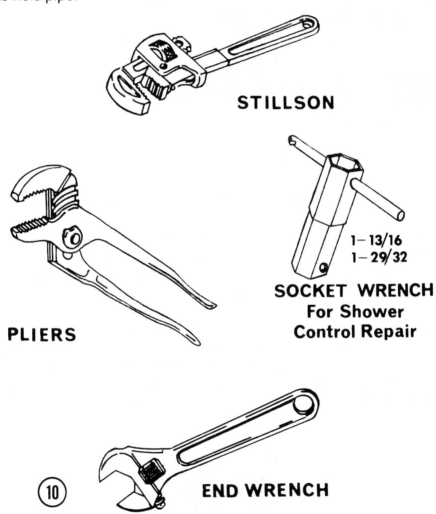

STILLSON

1— 13/16
1— 29/32

SOCKET WRENCH
For Shower
Control Repair

PLIERS

10 **END WRENCH**

Long handle pliers can sometimes be used in place of a stillson, if you are able to apply the strength required.

Two different size adjustable end wrenches are recommended. A roll of adhesive, or electric tape should also be kept handy. Wrap tape around a polished fitting before applying a wrench and you save the finish.

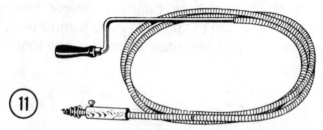

One snake, Illus. 11, is better than none, but two different sizes permit handling almost any emergency. The length of a snake depends on distance plumbing fixtures are from sewer or septic tank.

A set of socket wrenches, size indicated, Illus. 10, is needed when you replace washers in recessed shower controls.

If your kitchen is up to 45 ft. from a septic tank, a 50 ft. snake, Illus. 12, can frequently be required. While you may never need its full length, when it comes to plumbing, what you need in the way of tools is the least expensive way to make repairs. The cost, even of a 50 ft. snake, is frequently less than one plumber's visit. Snakes of all sizes can now be rented from many hardware and plumbing supply stores.

SNAKE or SEWER ROD

A short snake, also called a closet augur, Illus. 13, is ideal for opening clogged toilets. The toilet bowl is shaped as shown with some variation depending on manufacturer.

Before inserting augur in toilet,pull handle all the way up. Insert head in bowl, push rod down while you turn handle.

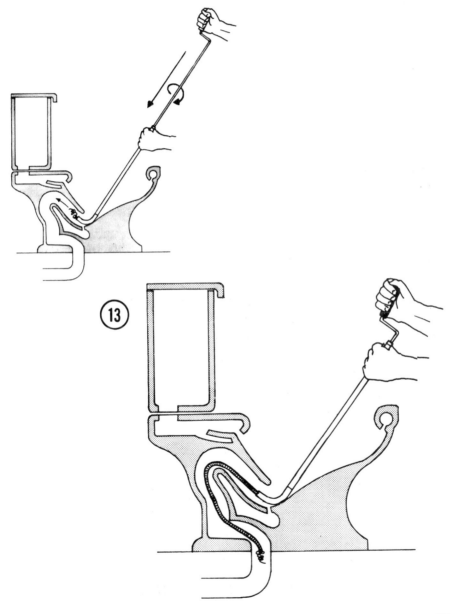

Using pliers, cut a wire coat hanger and straighten it out. Make an eye loop at one end, Illus. 14. You can frequently loosen and retrieve a comb, toothbrush, ball of paper, or other matter that never should have been dropped into a toilet. Keep turning the wire. Use care not to scratch finish.

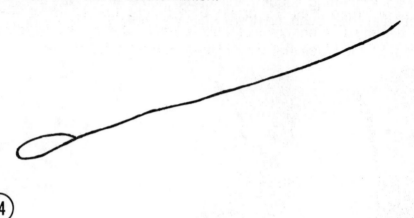

When a stoppage occurs in a toilet or sink, a plunger, Illus. 15, should be the first tool used. While it's only effective for minor stoppages, it will usually loosen up toilet paper, but not much more. First try a plunger, next a snake. If a snake doesn't open the line, open a cleanout plug, and again use a snake.

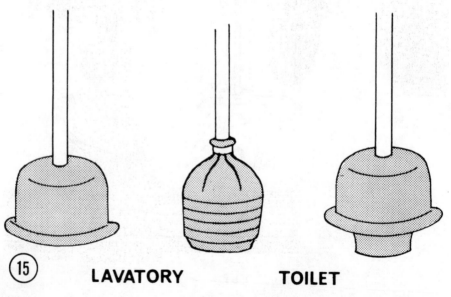

LAVATORY   **TOILET**

FAUCET REPAIRS

The first step in repairing a leaky faucet is to place stopper in drain, or cover drain with a rag to prevent screws and other small parts from disappearing down drain. Shut water off either below fixture, Illus. 16, or at water meter, Illus. 17. Since there are many different types of faucets and shutoff valves, directions explain two types that contain most component parts.

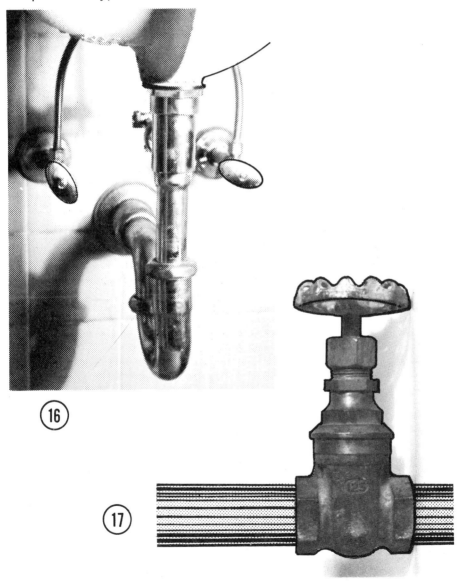

Remove cap in center of handle if your faucet has one, Illus. 3. These are either pressed or screwed on. If it's a press-on cap, use a small screwdriver to pry up. If it's a threaded cap, wrap edge with tape to prevent marring, then use pliers. Next remove screw holding handle.

Using an adjustable end wrench, remove packing nut. Replace handle on stem but not the screw or cap.

Turn stem, then lift out. Note screw and washer at bottom of stem. Remove worn washer. Clean cup or base before installing new washer. Replace with same size and kind of washer. If you can't get one immediately, you can sometimes get a little more use out of the old one by placing the good face down.

Look into faucet. Note whether seat is smooth. If there's any scale or grit, wrap the handle of a toothbrush with a rag and polish seat. Blow out loosened scale. If you see any cuts or notches on seat, ream seat lightly.

You can buy a reamer, Illus. 18, at hardware stores. Use reamer according to directions provided. Use care not to put too much pressure on reamer as the seat is soft.

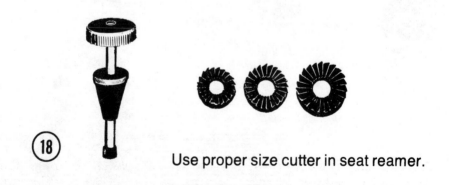

Use proper size cutter in seat reamer.

If seat can't be reamed smooth, replace with a new seat, Illus. 19. This usually requires a special wrench, Illus. 20. If one isn't readily available, use a slip on seat sleeve. These are pressed over present seat, Illus. 21.

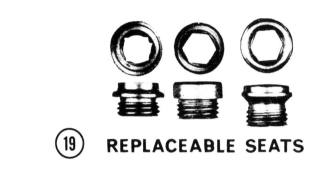

(19) REPLACEABLE SEATS

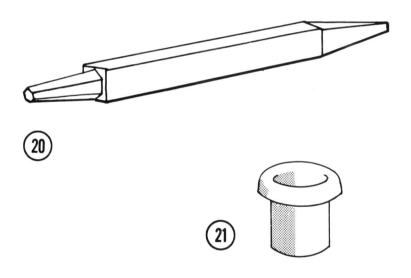

After replacing washer, fasten screw. Apply vaseline to threads on spindle, replace spindle. Replace washer under packing nut if stem had one when you took it apart, Illus. 3. Tighten packing nut.

If you can't get a packing nut washer, wrap stranded graphite asbestos wicking around spindle, Illus. 22. Turn packing nut down snug against wicking and it compacts into a watertight washer.

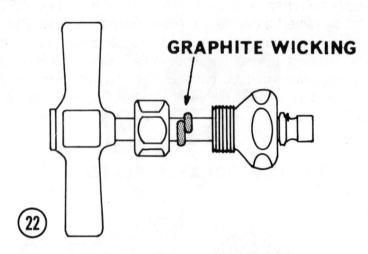

GRAPHITE WICKING

SHUTOFF VALVE

Illus. 23 shows the type of shutoff valve that sometimes needs repair. Here again you remove screw at top, remove handle, cap and packing nut. Remove spindle and replace washer at base of spindle, Illus. 3.

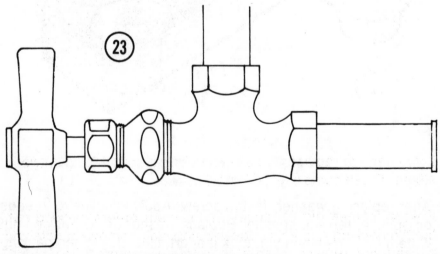

SHUTOFF VALVE

DECK FAUCETS

Handles on deck faucets, Illus. 24, 25, on sinks and lavatories, are usually fastened with a Phillips head screw, Illus. 26. Shut water off at valve below fixture. Open faucet halfway. Remove screw and gently pry up handle. Using a large end wrench, unscrew packing nut in same direction faucet handle turns. Keep turning stem while you loosen packing nut. Remove stem, packing nut and sleeve. Replace washer following procedure previously outlined.

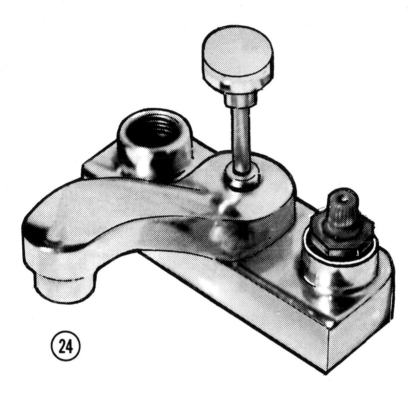

If you need to replace a desk faucet, measure spacing between stems. These come in 4", 6" and 8".

If water comes out of faucet where stem enters packing nut, tighten packing nut. If this doesn't stop leak, remove handle and packing nut, replace washer under nut or use graphite wicking, page 24.

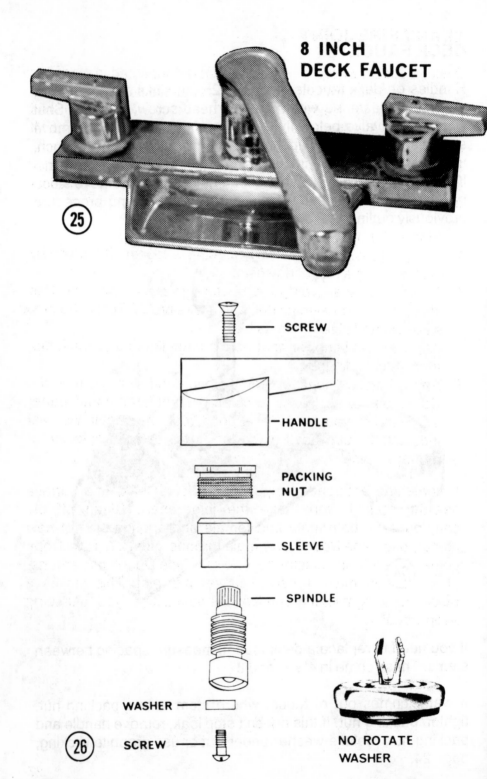

8 INCH DECK FAUCET

(25)

SCREW

HANDLE

PACKING NUT

SLEEVE

SPINDLE

WASHER

(26) SCREW

NO ROTATE WASHER

LEAKY PIPE JOINT

If a leak occurs where a valve or faucet is connected to pipe, first try to tighten joint with wrench. Use a stillson to hold pipe, an end wrench on valve. Never try tightening a valve using only an end wrench. Always support pipe with a stillson or pipe wrench. If tightening doesn't stop leak, you can try three other repairs. First, and easiest, is to apply Plastic Steel or equal quality compound with a putty knife. Follow this step-by-step procedure.

1. Shut water off to joint.
2. Relieve pressure in joint by opening a faucet. Drain line by also opening a faucet downstairs.
3. Clean rust, scale, dirt or other foreign matter off joint. Use steel wool or an electric drill with a wire brush. Polish the joint as bright and clean as possible.
4. Apply a heat lamp, or heat from a large light bulb, sun lamp, torch, etc. to dry joint.
5. Apply Plastic Steel, or equal compound with putty knife. Cover joint to thickness, and allow to set time manufacturer specifies. If you've done a good job, the repair will last indefinitely, but don't wait too long before replacing a damaged pipe.

If tightening a threaded pipe doesn't cure leak, and surface sealant doesn't work, take the joint apart. Clean off old compound on both male and female and apply a strip of new all-purpose Tape Dope® over male threads, Illus. 27. Tape Dope comes in a roll, like electric tape. Wrap Tape Dope once around, plus 1/2" on male thread. Reassemble joint. This makes a quick, easy, tight joint. Tape Dope is available at your plumbing supply dealer.

If pipe is 2" or larger in diameter, wrap Tape Dope at least twice around.

You can also repair a fine leak between a fitting and pipe, or in a cracked pipe or fitting, by soldering. Dry and clean a crack thoroughly as previously described. Apply Swif,® or equal solder, Illus. 28. Apply heat adjacent to crack. Do not apply flame on Swif solder. Use wire solder and flux to fill a large crack. Swif solder is easy to use. It contains flux. Brush Swif solder on following manufacturers directions.

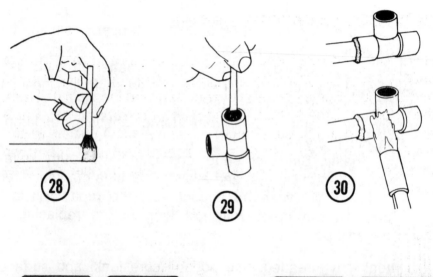

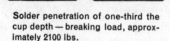

Solder penetration of one-third the cup depth — breaking load, approximately 2100 lbs.

Solder penetration of the entire cup depth—breaking load, approximately 7000 lbs.

If you find it necessary to remove a length of copper pipe, use a pipe cutter or hacksaw. Apply torch to fitting to loosen up a soldered joint. Cut replacement pipe to length required.

Clean ends of pipe and inside of socket fitting with steel wool. Be sure all particles of steel wool are wiped off.

Apply Swif or equal 50-50 tin lead solder and flux in paste form to end of pipe and to inside of socket, Illus. 29.

Join pipe and fitting, Illus. 30. Allow the small "collar" of solder paste to remain as shown. Apply torch to pipe and fitting but not directly on solder. When fitting and pipe heat up the solder turns from metal grey in color to black. It will now start bubbling. Remove flame and allow to cool. The heat draws the solder into joint. When bubbling stops, use a damp cloth to brush joint clean.

CLOGGED LAVATORY DRAINS

Hair, odds and other ends, that slip down a lavatory drain are usually caught by the stopper, Illus. 31. This should be removed and cleaned. Some can be removed by giving the top a half turn, then pull up. Others, Illus. 32, require unscrewing sleeve at A and pulling rod. Lift out plunger. Remove accumulated waste. Replace plunger so eye, Illus. 33, lines up with rod. Insert rod through eye. Tighten locking sleeve. Test plunger to make certain it's hooked up properly.

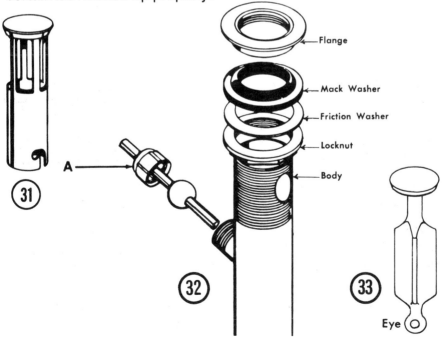

31

A

Flange

Mack Washer

Friction Washer

Locknut

Body

32

33

Eye

If the stopper doesn't free drain, the problem is probably in a clogged trap. First try a drain cleaner following manufacturer's directions. If this doesn't free up fixture, remove plunger and try a small snake, Illus. 34. Carefully turn handle on snake to turn head. You can work this kind of snake clear through trap.

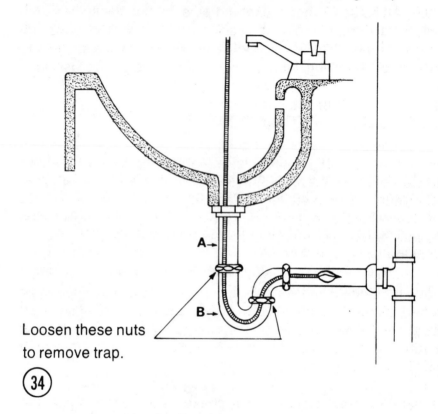

Loosen these nuts
to remove trap.

(34)

If sink or lavatory trap has a clean out plug at bottom, Illus. 49, remove plug. Be sure to place a large flat pan beneath trap to catch waste.

If you previously tried to open line with a drain cleaner, the water in trap is dangerous. Don't allow it to come in contact with your hands, arms, face, eyes or skin. Wear rubber gloves. Cover wrists and all exposed skin before opening trap. Open trap slowly to allow water to trickle out. Don't put your hand in water to retrieve a screw, plug, or wrench. Don't throw this water on the lawn or plants.

If there's no clean out plug at bottom of trap, and you can't get a snake through trap, (this is a very unlikely situation), remove trap completely by loosening packing nut, Illus. 34.

To loosen, turn packing nut to right. When reassembling, insert A into B as far as it will go, then start fastening packing nut. Always replace washers rather than reuse old ones. Buy the same kind and size. If they have a beveled face, be sure to replace in exact position.

CLOGGED SINK DRAIN

Before leaving a house for three or more days, flush sink with lots and lots of hot water. This will loosen up and wash down particles of food, grease, etc. Then, to play safe, use a good drain cleaner. But make certain all cleaner is flushed out of trap and lines before departure. Many of even the best drain cleaners harden up when left longer then directions specify, so use plenty of hot water.

If your sink does clog, use care in selecting a drain cleaner. Most contain dangerous chemicals. Read and follow directions on can. Never allow water or any dampness to get into container. If you put a cap on a can containing only a few drops of water, it could blow up in your face.

If a plunger, drain cleaner, or a snake hasn't opened line, the next step is to open up the nearest drain plug, Illus. 8. Remove plug, run snake through line. If plug is rusted, use penetrating oil as suggested previously. Give oil time to penetrate, then tap lightly with a hammer to jar rust.

Always open cleanout plug nearest fixture. If the first cleanout plug doesn't open the line, try the next one, Illus. 8.

Always use an electric heater or heat lamp to warm up a trap. Traps, insulated against room heat by under sink enclosures, frequently clog in severe cold weather. In cold weather always keep door of a sink enclosure open.

Never use drain cleaner if sink is equipped with a garbage disposal unit. Use snake through trap. A pound of washing soda used weekly will keep most disposal drains open.

CLOGGED TOILET

First try a plunger. If this doesn't clear drain, use a short snake in bowl, Illus. 13. If this doesn't clear obstruction, locate nearest cleanout plug in main waste line, and use a long snake, Illus. 12.

TOILET, LAVATORY, SINK REPAIRS

When you buy or build a house, make a record of all equipment; the name of manufacturer, year and model number, and where possible, the serial number. Immediately purchase a set of replacement washers for all fixtures. Identify each so you know where each is used. While you may not need any of these until they are too dried out to use, having samples of what you need, when you need it, makes life worth living. It also insures your making a professional repair in the shortest possible time. Being a genius just requires a little foresight.

Always identify whether a real old toilet is one piece or two piece. Also measure spacing of seat bolts, center to center. Do they measure 5-1/2", 7", 7-1/2" apart? This can be an important clue to its actual model number.

Note whether seat bolt (frequently called seat posts) go through front of water tank, or into top of bowl. Also note whether flush handle is on front of water tank, or on side.

Note whether bowl is fastened to floor with four bolts, or two. The china knobs, Illus. 35, covering nuts, or covering plates, Illus. 36, indicate how many bolts hold toilet to flange.

Only with definite information as to model number, can your plumbing supply house substitute an accurate replacement, if he can't replace the manufacturer's original parts.

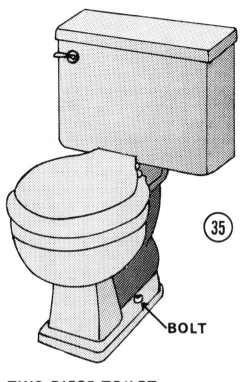

TWO PIECE TOILET

BOLT

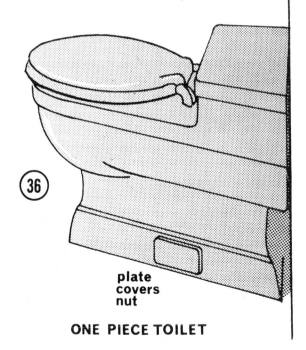

plate
covers
nut

ONE PIECE TOILET

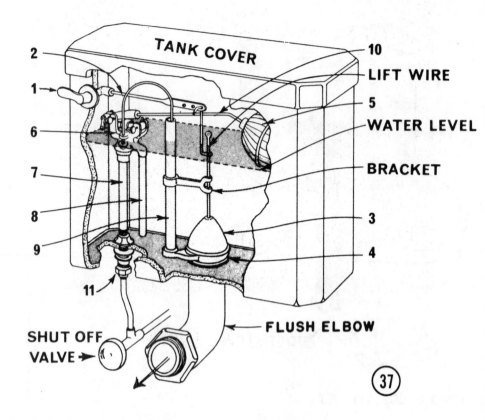

2 TANK COVER 10
LIFT WIRE
1
5
6 WATER LEVEL
BRACKET
7
8 3
9 4
11
SHUT OFF
VALVE FLUSH ELBOW

(37)

A toilet operates in the following manner. When you press handle 1, Illus. 37, it raises lever 2, which in turn lifts Tank Ball 3. Water in tank rushes down open valve 4, pours into bowl into waste line, either to septic tank or city sewerage system.

TANK ON TWO PIECE TOILETS

When water in tank flushes down toilet, the float 5 drops. This releases pressure on ballcock 6. Water, coming into tank through riser 7 flows through ballcock into hush tube 8. When tank ball closes valve 4, water raises float. When float reaches its specified height, it applies pressure and closes ballcock.

Everytime a new house is connected to the city water line, the water line picks up a lot of foreign matter. If dirt or sand gets in, it could lodge against ballcock washer 6, or on tank ball seat 4. This allows a trickle of water to flow into tank or bowl, Illus. 40.

If water continues to flow into tank, it could be caused by many different reasons, and it can be stopped by any one.

1. Make certain handle, Illus. 38, isn't stuck in open position. The tank cover might be pressing against handle, holding it down.

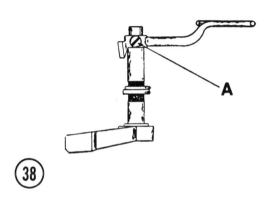

2. Screw A, Illus. 38, holding handle, might have loosened up and allowed stem to move out of position. Make certain handle is free and works properly.
3. Check to see if tank ball is on valve. If tank ball doesn't fall automatically into position on valve, it could be caused by rust on lift wire or in bracket. Remove lift wire and steel wool shaft clean. Apply vaseline to lubricate holes through bracket.
4. If tank ball is on valve, and water continues to flow, shut water off, flush toilet. Remove tank ball by unscrewing lift wire. If tank ball is worn, shows ridges, or rust buildup, replace ball.
5. Inspect flush valve seat 4, Illus. 39. Use a rag to polish rim. Remove rust, dirt, etc. A particle of sand on edge can keep tank ball from seating properly. After installing new tank ball turn water on to see if it shuts off automatically at proper water level. If water continues to run into overflow 9, Illus. 37, check to see whether float 5 is operable.

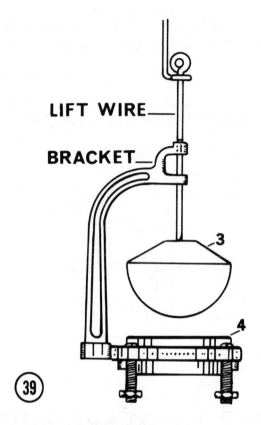

LIFT WIRE

BRACKET

3

4

(39)

6. If float 5, is operable. Again shut water off, flush toilet, unscrew float. Shake float to see if it contains any water. If you hear any water in float, replace float.

7. If water continues to run, bend float arm 10 down slightly. This should automatically shut water off at a level lower than that indicated on inside of tank.

8. If water continues to run, again shut off water, flush toilet and remove ballcock, Illus. 40. This is done by loosening thumb screws 1 and 2. Slide lever out, remove ballcock, Illus. 41.

To replace ballcock assembly, Illus. 40, disconnect coupling nut #11, Illus. 37. Remove float and rod #10. Disconnect lock nut #4, Illus. 40; lift out assembly. Install replacement making certain it has the same length shank #5, Illus. 40. Using pliers, Illus. 10, tighten lock nut only enough to compress and make shank washer #3, Illus. 40 watertight. Use extreme care as too much pressure could crack fixture.

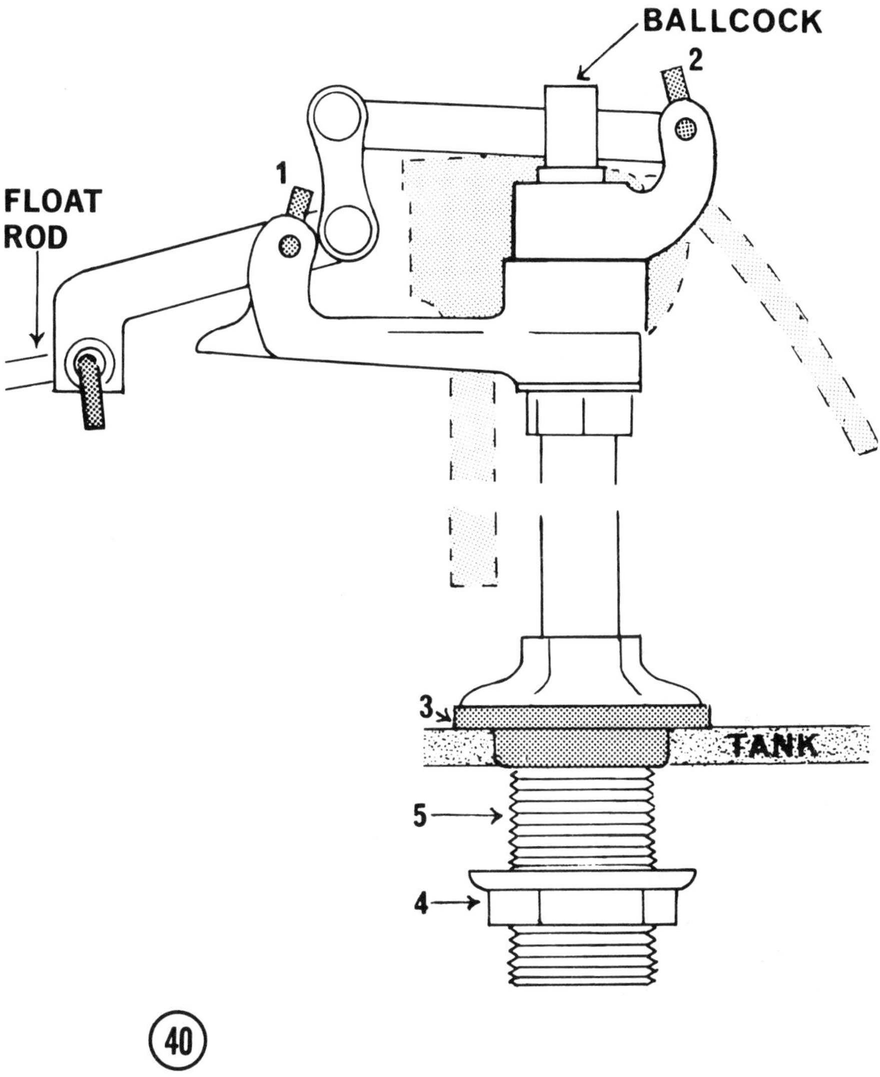
BALLCOCK
2
1
FLOAT
ROD
3
TANK
5
4
40

Since there are many different ballcocks in service, yours may differ. It should still have most of the same working parts. When you know "which does what," you make necessary adjustments. A good investment, even before you need it, is a ballcock repair kit that fits your ballcock.

To replace a washer on a ballcock, Illus. 41, 42, loosen screw at bottom and replace washer with exact size.

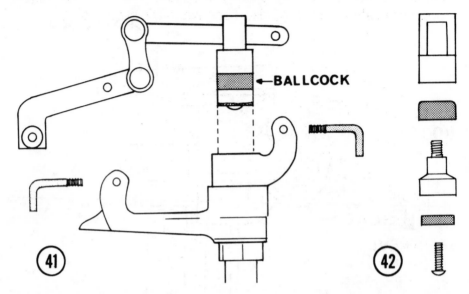

If ballcock has a ring nut on bottom, Illus. 43, use care not to burr nut when removing. To protect, wrap edge with tape. Take worn washer to store to make certain you obtain exact kind and size. If it's not possible to obtain washer required, turn worn washer over, place good face down, but by all means, get a replacement as soon as possible.

Replace split washers, Illus. 44 with same kind. Look down into seat of ballcock to see if there's any rust, sand or particles of dirt cemented to seat. Clean thoroughly, apply vaseline and replace. Again test toilet.

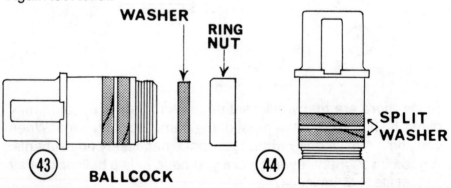

If water coming into tank creates loud noises, you can frequently eliminate noise by decreasing pressure. To do this adjust pressure valve B, Illus. 45. Replace washer if required.

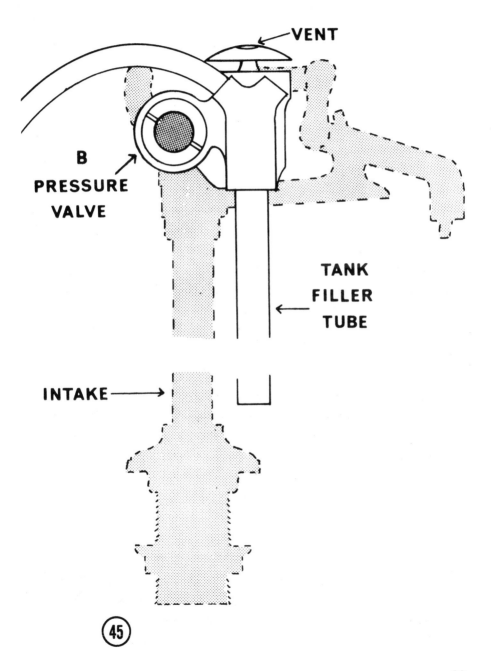

VENT

B
PRESSURE
VALVE

TANK
FILLER
TUBE

INTAKE →

45

If water spouts out of vent, Illus. 46, shut water off, flush toilet. Loosen screw at top. Remove cap and unscrew vent A. Clean rust, replace washer. Also clean rust in chamber. Apply vaseline and replace.

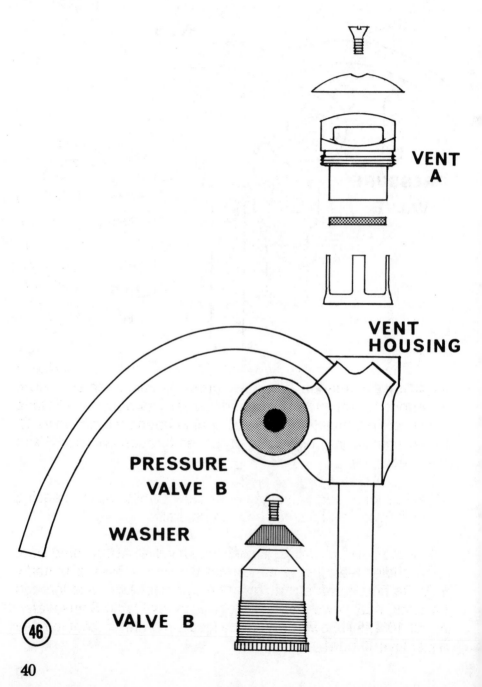

VENT A

VENT HOUSING

PRESSURE VALVE B

WASHER

VALVE B

46

Illus. 47 shows a rim type pressure valve. Screw A, holds ring in position. Loosen A and turn knurled rim B clockwise to cut down pressure.

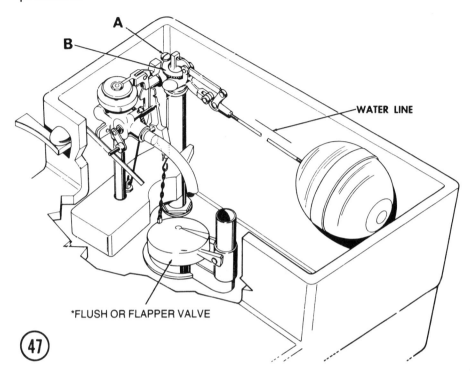

(47)

Most one piece, no-overflow water closets, Illus. 48, operate in the following manner. When you press handle, it remains down for about 10 seconds. The handle lever lifts tank ball allowing water to rush down flush valve*; it also opens the rim valve C. Water coming into riser tube D rushes through ballcock E into rim of bowl.

This flushes and cleans side of bowl while water passing through flush valve flushes bowl down drain.

When the tank ball drops, it closes flush valve. At the same time, the handle returning to its normal position, closes a butterfly valve to rim. Water still rushing through riser tube, and through ballcock, now flows into hush tube F to refill tank. Some water, about 10%, still flows through ballcock into rim of toilet to raise water level in bowl.

41

When water raises float to proper level, it closes ballcock. The hard rubber ball in vent G, Illus. 48, drops into position and prevents water from siphoning from tank into bowl.

If incoming water spurts out of Vent Cap and leaks out of tank cover, adjust water pressure at B. Also inspect and replace rubber washer in Vent Cap, replace Vent Ball. If you can't get a replacement ball and washer, put an empty aluminum foil cheese container over cap in position shown. While it should stay in place, a dab of contact cement on inside of tank cover, or on bottom of container, should keep it in exact position.

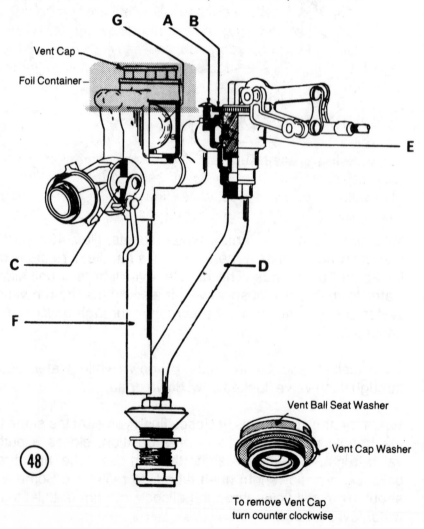

Vent Cap

Foil Container

G   A   B

E

C

D

F

48

Vent Ball Seat Washer

Vent Cap Washer

To remove Vent Cap
turn counter clockwise

42

CRACKED TOILET TANK OR BOWL

If a toilet tank is damaged, and the break leaks water, you can frequently make a temporary repair using a silicone tub caulk now on sale in most hardware stores. Shut off water to fixture. Drain and sponge dry. Thoroughly clean area around crack before applying sealant. Apply sealant exactly as directions specify. Be sure fixture is in a warm room. Don't attempt to repair a cold fixture with a sealant that requires 70° to set up.

If fixture has a through break, or large particles have come loose, apply tub caulk to outside, allow to harden time directions prescribe before applying tape reinforcing. After applying tape on outside of fixture, apply tub caulk to inside face of crack. Allow to set time manufacturer prescribes before turning on water.

FIXTURE REPLACEMENT

If fixture is beyond repair and needs replacement, buy an exact size replacement. Shut water off at shutoff valve. Place open pan below and disconnect water line at A, Illus. 49. Disconnect waste line at B, loosen nut. Lavatory is fastened to wall at C.

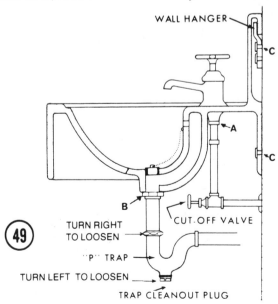

WALL HANGER

•C

•A

•C

49

TURN RIGHT
TO LOOSEN

CUT-OFF VALVE

B

"P" TRAP

TURN LEFT TO LOOSEN

TRAP CLEANOUT PLUG

HOW TO INSTALL A LAVATORY

Lavatories can be fastened to the wall by hanging top edge over wall bracket C, Illus. 49, and/or, through fixture at lower C, or to legs. If you want to hang it to the wall, use brackets that come with lavatory. Always install a replacement lavatory at height that permits connecting present water and waste lines.

LOOSE TOILET BOWL

Through misuse, juvenile delinquency, or poor installation, toilet bowls do loosen up. If inspection merely reveals a loose bowl, no water seepage at base, you can frequently take up slack by removing the porcelain caps, Illus. 50. These are usually secured with caulking. Older installations used plaster of paris. Using a knife and gently tapping with hammer loosens plaster without cracking cap. The nuts holding toilet to flange, Illus. 51, will then be revealed. Try tightening nut but not too much or you will crack fixture.

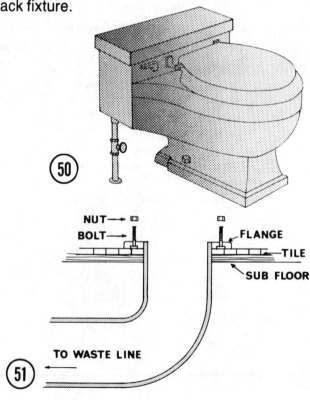

NUT— ▭
BOLT—

FLANGE
TILE
SUB FLOOR

TO WASTE LINE

If water is leaking at base of a one piece toilet, shut water off, flush toilet, sponge out water, disconnect water line to tank. Loosen and remove nuts, lift toilet straight up. Scrape away old seal. Buy a new seal, Illus. 52, from your plumbing supply house.

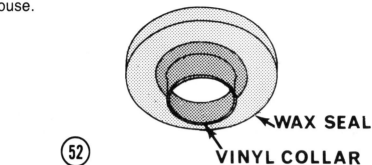

WAX SEAL

(52) **VINYL COLLAR**

HOW TO REPLACE TOILET BOWL SEAL

To replace a seal, Illus. 52, apply setting compound, available from your plumbing supply house, Illus. 53. Place new seal in position on horn of toilet and lower toilet directly over flange on floor. Don't come in at an angle. Don't slide it into place. Come straight down. Press toilet down with a slight twisting motion. Be sure to aim toilet so bolts in flange come through holes in base. Fasten nuts snug, but do not tighten with force. Apply setting compound around nut and replace caps. Reconnect water line using Tape Dope on male threads.

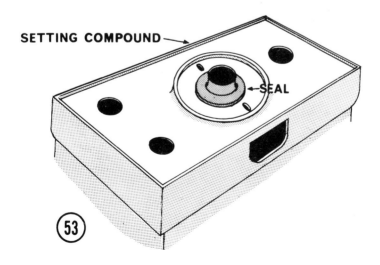

SETTING COMPOUND

SEAL

(53)

If a new tile floor prevents bowl from seating against floor flange, use two wax seals, one on top of the other, to insure a tight joint.

When reconnecting water line, use care not to twist ballcock assembly from its original position.

If water leaks out at base of a two piece toilet, disconnect flush elbow, Illus. 37. Loosen and remove nuts, lift toilet straight up and place it bottom side up. Carefully handle unit. Scrape away setting compound on edge of toilet, also remove old seal.

Place a level on flange and on floor where toilet sets. If floor has settled, it may be necessary to shim toilet. Use small pieces of wood shingle placed close to bolts. Replace toilet and check with level. If it checks OK, carefully lift toilet out of position without disturbing shims. Lay a thin bead of plaster of paris on rim. Use a thicker bead alongside a shim. Replace toilet seal. Press toilet into seal and check with level. Only tighten bolts snug. Use extreme care not to crack fixture when tightening bolts. Remove excess plaster of paris. Clean edge of rim with a damp rag.

LOOSE TOILET SEAT

This can usually be traced to a worn rubber washer, Illus. 54, or worn rubber snubbers, Illus. 55.

64

Remove tank cover, shut off water, flush toilet. Use end wrench to loosen nuts. Remove seat and bolts. Replace washers on shaft with beveled face facing tank. Insert bolts in holes, apply washers and nuts. Tighten snug but do not force.

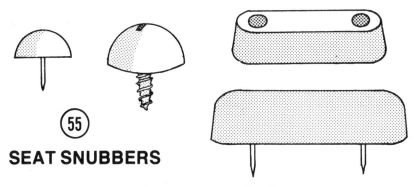

SEAT SNUBBERS

These are made with screws or nails. If nail type, use pliers, pull out worn snubbers, replace with new ones, but not in the same holes. Use same size snubbers.

REPLACE TOILET SEAT

Measure spacing between seat bolts and buy replacement that fastens in same position. If present seat bolts go through tank, be sure replacement bolts fits this installation.

It isn't necessary to shut off water, or flush toilet, but since there's a good chance of dropping a nut, washer or wrench into tank, shut water off, and flush toilet.

If nuts holding bolts have rusted, use liquid wrench to loosen. Allow oil to penetrate. Tap nut lightly and it now should turn freely. Don't tighten nuts with force. You will crush rubber with too much pressure, thus shorten its life span.

BATHTUB FAUCET REPAIRS

Shut water off at meter. To replace seat, or washers in this late model bathtub and shower control, Illus. 56, 57, pry or screw off cap A, remove screw B, knob C, and escutcheon plate D. Stem bushing E fits over stem F. Place knob back on stem and remove stem. The O ring can then be replaced. Renewable seat G can also be replaced if need arises. Some stems are held in place with a packing nut and washer, Illus. 58. Follow procedure outlined for replacing washers.

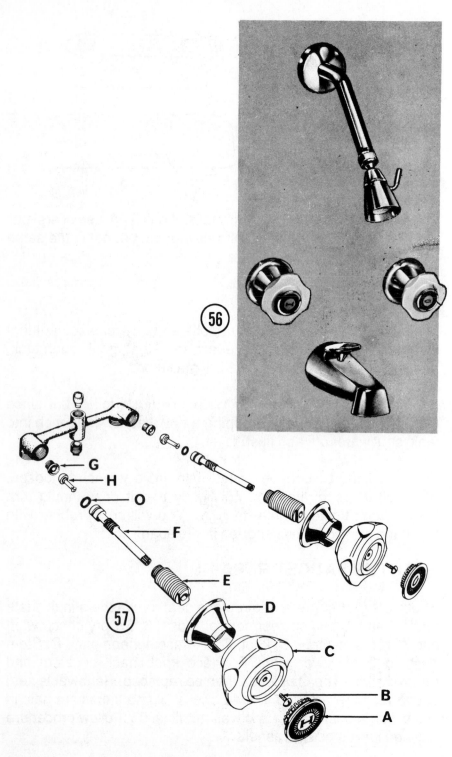

G

H

O

F

E

56

57

D

C

B

A

48

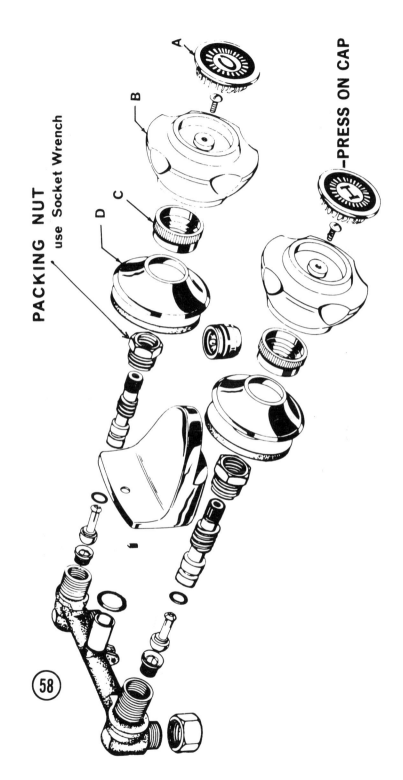

PACKING NUT
use Socket Wrench

A

B

C

D

PRESS ON CAP

58

49

BATHTUB DRAIN

You can remove stoppage from some bathtub drains by removing drain plate A, Illus. 59, 60. The plunger assembly B can be removed by loosening screws C. Lift plate and plunger out of pipe. A small snake can be worked through drain and trap. Most reliable builders provide an access panel that permits working on bathtub drain and trap. This is usually screwed or hinged on other side of wall on drainage end of tub. The tub trap can be serviced following same procedure for a sink trap.

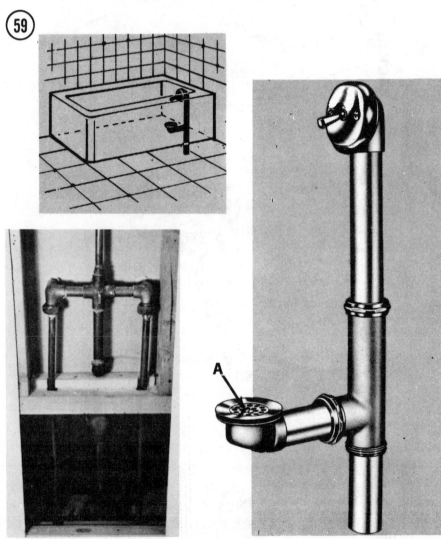

(59)

A

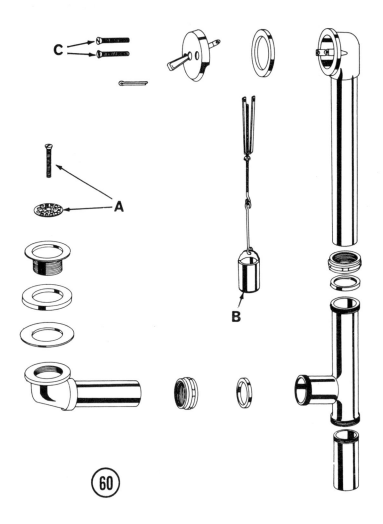

SUDDEN TOILET STOPPAGE

If a toilet begins to clog, or slow down carrying waste out of the bowl, the trouble might be traced to a change in toilet paper, Kleenex, etc. Soft tissues frequently stick to sides of a bowl, trap, or waste line, while coarse paper flushes down.

LOST & FOUND DEPARTMENT

If anything valuable falls down a lavatory or sink drain, or toilet bowl, don't run water or flush toilet. Search a toilet bowl as far as your hand will go before opening plug in waste line.

If you don't find it in the lavatory or sink trap, open trap in main waste line, Illus. 61. First open A. Catch water in a heavy polyethylene bag or clean garbage can. If this doesn't produce results, open B and C. If you still don't find it, close C, open A. Flush trap through B with a garden hose. If this doesn't locate what you're seeking, it may be caught in drain line above. Close A and C finger tight. Plug line to sewer at B by stuffing a big rag in opening. Half fill bath tub with water and on signal, have someone open tub drain and flush a toilet. With waste line full, open plug A, water rushing through drain and trap should dislodge any small object while it fills several barrels with water in your basement. If this doesn't locate the missing object, file a claim with your insurance agent.

61

KITCHEN SINK AERATOR

Here's what to look for when you want to clean a strainer in an aerator. Unscrew A, Illus. 62, 63, and clean strainer. Some aerators have a stream diffuser B that frequently collects grit or sand. If water leaks out between A and C, replace washer D.

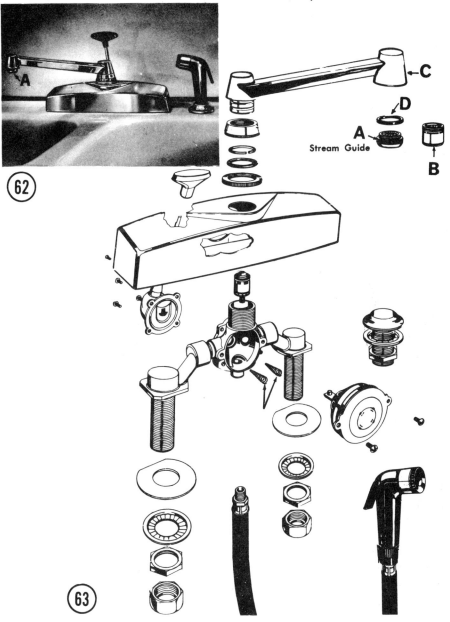

Stream Guide

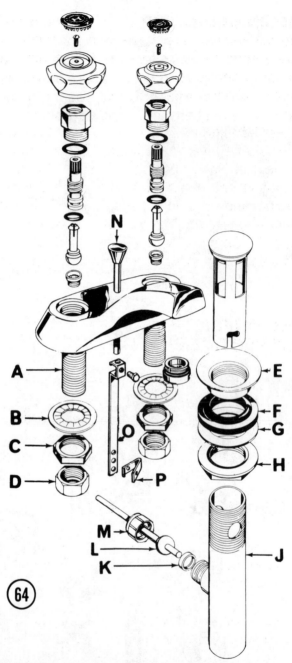

Illus. 64 shows how many lavatory faucets are fastened in place. The threaded shank of fixture A is inserted through opening in lavatory. Shank washer B is placed snug against bottom surface. Locknut C holds fixture in position.

Coupling nut D slips on flanged water line and is fastened to A. While some connections are made with a washer, many fixture manufacturers have eliminated same. Follow manufacturer's, or plumbing supply dealer recommendations as to when to use sealing compound and/or washer.

Most installers bed flange E for waste line in setting compound. The mack washer F, is placed under fixture with beveled face up. This is held in place by friction washer G and a locknut H. Screw locknut H onto drain J. Screw drain J into E. Tighten locknut H. The pop-up waste lever N slips into rod O and is fastened with a setscrew. The ball rod L is fastened to O with a spring clamp P. Coupling M fastens L to J. Be sure beveled washer K is placed in position to receive ball on rod. Waste line J connects to trap, Illus. 49.

PLUMBING CHECK LIST
FOR PROSPECTIVE BUYERS

The plumbing represents a big factor in evaluating the cost of a new home. The existing plumbing represents 10%, or more, of the price you pay for a house. It could represent as much as 20% if you have to replace it. For this reason, don't hesitate to spend time studying the condition of pipes and fixtures.

WHAT TO LOOK FOR

Note all exposed pipe joints. See if there's any discolorization caused by water leakage on walls or floor. Note whether there's any rust, moisture or leakage.

Turn each faucet on in kitchen and bathroom to see if you have as much pressure upstairs as down. See if there's enough pressure to have a downstairs faucet open while you draw water in an upstairs bathroom.

Taste the water. Is it to your liking? Note whether there is any odor, rust, scale, corrosion, or too much chlorine.

Flush each toilet. See how quickly water goes down drain. Flush it a second and third time to make certain there's no stoppage in a long waste line. Does the toilet work quietly, shut off as it should?

Check each sink, lavatory and tub. Place stopper in drain. Fill fixture and note how long it takes water to drain off.

Try the hot water line. Allow it to run a while to see how much hot water the system provides; how long it takes to "recover." Hot water heaters and tanks in many older homes were installed prior to dishwashers and washing machines. Many don't have the capacity to handle this equipment. In many cases the addition of only one piece of equipment requires a larger hot water tank.

If the house is served by a well, find out who has been servicing it. Get a written statement from owner concerning past water supply, and what he thinks its capacity is. Many wells fluctuate in capacity due to seasonal rains. The best time to judge a well's capacity is during, or after, a long drought, or at the end of a hot summer.

Are waste line clean out plugs easy to service? These should be placed in position shown, Illus. 61. Cleanout plugs should be located at least every 40 ft., never more than 50 ft. apart. Does a hot water heating system have an easy to reach drainage plug. Are waste lines to sink and lavatory undersize or do they meet local codes? A sink waste line should be 1-1/2" minimum. If line is longer than 40', it should be 2" minimum.

Does the water line chatter when you close a faucet quickly? Does a hot water heating line set up noises when supplying certain radiators?

If the house is heated with hot water lines embedded in a concrete slab, do rugs, or newly laid linoleum, or tile, cover a bad crack in concrete, or a ruptured pipe? It's always well to insist on turning the heat on, even during mid-summer, to test the system for noise, leaks, etc. Some boiler circulators, and/or pipes leak only occur when the system is operative.

Go up into attic and see whether there is, or was any leak around vent pipes going through roof. Water stains on framing, roof rafters, attic floors, etc., provide signs of previous leaks.

FROST FREE OUTSIDE SILLCOCK

If winter arrives before you have shut off an outside faucet, a frost free, self-draining sillcock, Illus. 65, will eliminate a freeze-up. Always remember to disconnect and drain a garden hose.

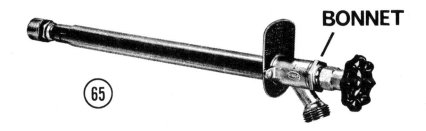

BONNET

Select sillcock length required, Illus. 66, to project through wall, so valve seat, Illus. 66A, is within a heated area. Sillcocks come in 6", 8", 10" and 12" lengths. Drill a 1" hole through foundation. The sillcock shown in Illus. 65, is available with a Universal connection, Illus. 66B. Threads A take a 3/4" M.S.P.S. Threads B take a 1/2" F.S.P.S. If you remove the Universal adapter A., you can sweat an adapter, Illus. 66C.

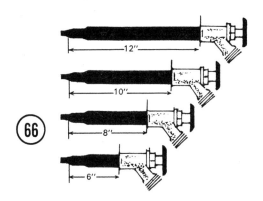

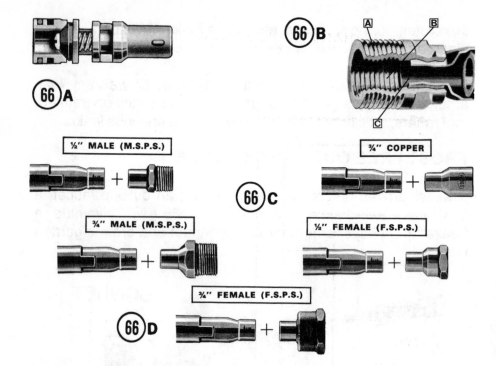

½" MALE (M.S.P.S.)

¾" COPPER

¾" MALE (M.S.P.S.)

½" FEMALE (F.S.P.S.)

¾" FEMALE (F.S.P.S.)

Before installing sillcock, unscrew bonnet, Illus. 65, pull out assembly. Always replace assembly AFTER sweating connection to pipe.

If you want to connect copper tube to equipment having standard iron pipe connections, copper fittings are available with either male or female threaded ends, Illus. 66D.

FACTS ABOUT FITTINGS

Threaded fittings are designated F.S.P.S.—Female Standard Pipe Size; or M.S.P.S.—Male Standard Pipe Size. All fittings are either male or female, threaded or sweat. In the case of plastic, you substitute solvent for solder.

HANDY SIZING SCALE

To determine size of waste and water lines, wrap a strip of paper around pipe and mark paper where it begins to overlap. Place strip against scale to ascertain size of pipe, Illus. 67.

HANDY PIPE SIZING SCALE

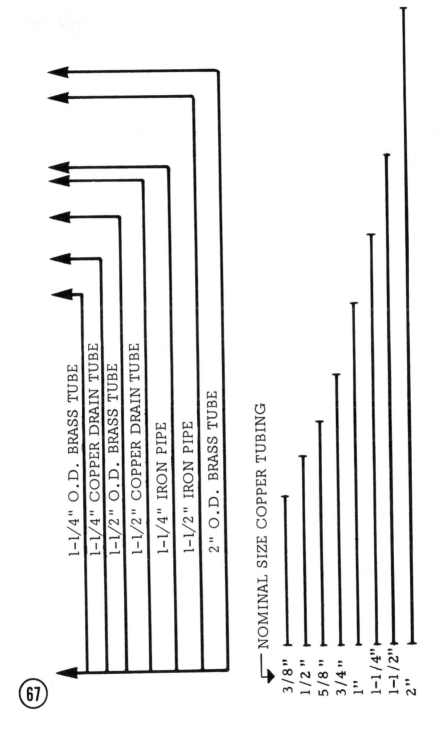

1-1/4" O.D. BRASS TUBE

1-1/4" COPPER DRAIN TUBE

1-1/2" O.D. BRASS TUBE

1-1/2" COPPER DRAIN TUBE

1-1/4" IRON PIPE

1-1/2" IRON PIPE

2" O.D. BRASS TUBE

NOMINAL SIZE COPPER TUBING

3/8"
1/2"
5/8"
3/4"
1"
1-1/4"
1-1/2"
2"

67

THAWING FROZEN PIPES

Electric heating cables, Illus. 68, are available in various lengths up to 100 ft. Wrap these around exposed pipe, plug into any outlet and in most cases your prayers will be answered. Never overlap cable or allow edges to butt against each other. Many cables come with thermostat controls. These can be plugged in and left connected throughout the winter.

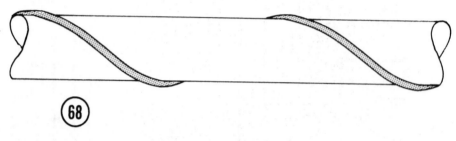

(68)

Pipes running through an unheated crawl space frequently freeze when heat in room above drops below normal temperature. Homeowners who install insulation between floor joists over crawl space frequently fail to protect pipes. Since the heat loss is lessened, pipes need more protection. Protecting hot and cold water pipe with a thermostatically controlled tape is a sound investment.

SUMP PUMP

If water collects in your basement, a sump pump can relieve the situation. Check your leaders to make certain they are carrying water away from your house and not draining into footings.

If water collects in one corner of your basement, the situation can be corrected by digging a 12" diameter hole 12", 18" to 24" deep. Use a chisel and hammer to break through concrete or rent an electric hammer. Since these work fast, rent one by the hour. Cut hole to size and depth sump pump manufacturer recommends. Most pump manufacturers recommend filling bottom of hole with 3" to 4" of gravel, then placing an 8" to 12" diameter drain tile vertically in position, Illus. 69.

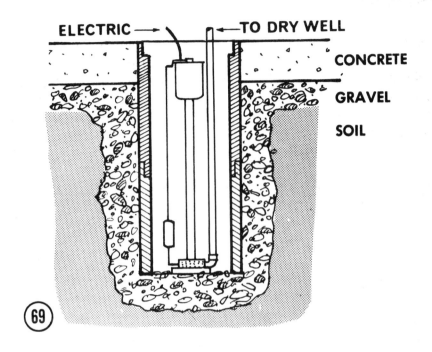

ELECTRIC → ← TO DRY WELL

CONCRETE

GRAVEL

SOIL

69

Drain tiles designed for sump pump installation are perforated to allow for seepage. Back fill around tile with 3" to 4" of 1" to 1-1/2" crushed stone. If you can't buy a cap for the tile, make one from 3/4" exterior grade plywood. Cut and glue two thicknesses to make a 1-1/2" thick cap. Cut discs to inside diameter of tile, A, Illus. 69. Top of tile should finish flush when you repair floor.

Drill holes in cap, one for pipe, another for electric conduit, plus several more to allow circulation of air to pump. Connect water line from sump pump to a dry well. Since you will have to go through foundation wall to discharge water, drill hole through foundation at highest possible point. You can use plastic pipe. When buying pump, be sure to find out whether it will lift water to height your basement wall requires.

If your basement is damp, install a humidifier. Select a size with sufficient capacity to handle overall size cubage of your basement.

HOUSEHOLD HINTS

To increase disintegration of solids in a septic tank, and to stimulate bacteria growth, add yeast/or sugar to a septic tank every two months. This stimulates growth frequently destroyed by harsh detergents. Mix yeast in water and flush down a toilet or sink.

When closing a house during winter, to eliminate damage from freeze-ups, always shut off main water line valve, open all faucets, open and drain all valves. To eliminate any freeze-ups, pour anti-freeze or denatured alcohol into every trap and toilet bowl. Sponge all water out of toilet tanks. Drain heating lines and boiler.

Always ascertain name of manufacturer and catalog number of all plumbing and heating equipment in your home. Make a reference guide showing shut-off valves and what they control.

PLUMBING WALL

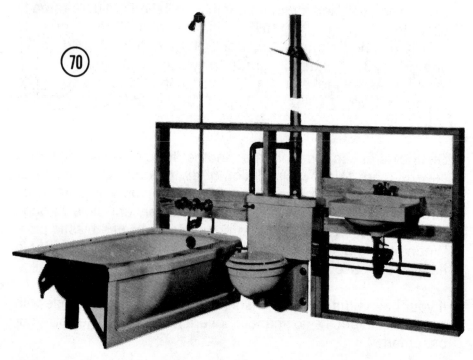

For those who want to install a bathroom quickly, and with as little labor as possible, the EB Plumbing Wall, Illus. 70, provides an excellent solution.

No floor flange or closet bend is required when installing the EB Plumbing Wall, Illus. 70. Toilet waste empties into wall inlet, as does drainage from lavatory and bathtub. Since the plumbing wall is easy to install, select any location you desire. It is especially suited to one story houses built on a slab, and in two story installations when you don't want to disturb a ceiling below.

WASHING MACHINE PROBLEMS

Problem: No water entering machine.
Solution: Check to see if someone shut off valve to washing machine line.
Check to see if water pressure is OK at other faucets.
If OK, the probable fault could be:
1. A clogged strainer, Illus. 71.
2. A loose washer on end of faucet spindle, Illus. 26.

When new homes are being erected in a neighborhood, mud and other foreign sediment frequently builds up. Remove hose from faucet and remove strainer. If clogged, replace. Your hardware store has replacements. If strainer is OK, or replaced, and the problem still exists, check washer on end of spindle.

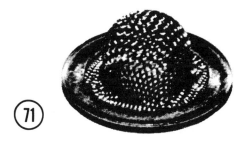

FAUCET REPAIRS SIMPLIFIED

When a faucet fails to provide water when turned on, the problem can usually be traced to a faulty washer. Shut off valve serving line. Remove faucet spindle, Illus. 26. If washer is loose or remains in faucet, you have discovered the cause. If washer remains in faucet, use thin, long nosed pliers to pull washer out. Replace washer and screw.

If screw head has rusted, removing shank of screw, and/or obtaining a replacement screw, may be difficult. Be sure to take stem to your hardware or plumbing supply store to insure getting proper size threads.

If same can't be obtained, as happens most of the time, use a No-Rotate, or equal swivel head washer, Illus. 26.

First select size of No-Rotate washer your spindle requires, then press stem into position.

If you have difficulty removing shank of a screw, spray or soak end of spindle in Liquid Wrench. After allowing oil to penetrate, tap shank lightly and turn with pliers, or use a No-Rotate washer.

No-Rotate washers lock into position and provide a simple solution to a troublesome problem.

WATER TANK REPAIRS

While pin holes in a galvanized water tank indicate its life span is nearing an end, you can make a quick repair by screwing in a self-tapping plug, Illus. 72. These come in various sizes. Always buy at least three different sizes. Use the smallest size first. Using an adjustable end wrench, just screw it in. If the spot is weak around the edges, the plug won't seat itself. Use the next size plug until you seat a plug securely in position. Keep these plugs handy as you can make an instant repair. This will give you ample time to shop for a replacement tank.

An ace plug, Illus. 73, simplifies repairing a larger hole.

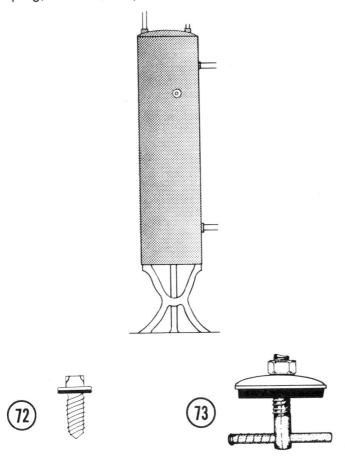

PLUMBING FACTS OF LIFE

Plumbers, like most skilled craftsmen, and other self employed —doctors, dentists and lawyers, frequently charge whatever they believe the customer can afford. Adjusting a faulty toilet, opening a stopped up drain, drilling and filling a tooth, or prescribing a remedy, have one common denominator, each requires the sale of time and experience. Each segment of society earns its living rendering service to others. What you pay depends entirely on your knowledge and willingness to do the work yourself. Keep well and you save doctors bills. Keep plumbing in repair and you effect comparable savings.

HOT WATER HEATING PROBLEMS

If you consider buying a house that has hot water heating in a downstairs playroom, or in any room where pipes may have been embedded in a concrete slab, check the water pressure boiler gauge, Illus. 74, after water has been heated to required temperature.

(74)

When all radiators are full, and boiler is at proper pressure, shut the valve on the line supplying the automatic feed valve, Illus. 75. Turn the thermostat up so the circulator starts pumping water through the system.

A hot water pipe, buried in concrete, can spring a leak and still permit the system to operate satisfactorily. While an automatic intake valve will normally keep the pressure up, the burner will run longer, and at more frequent intervals. A leaking system can be operative with no one knowing a leak exists.

Since most hot water heating supply and return lines, Illus. 76, bedded in a concrete slab, are usually laid over a filled-in area, sometimes covered by #15 felt, sub and finished flooring, water frequently sinks without appearing on the surface. Close the intake supply line and the boiler gauge will drop. Unless you have a map showing exactly where the supply and return lines are buried, a repair can be a costly and frustrating job.

The hot water line from boiler is connected to a circulator. The circulator supplies radiators through a main supply line plus feeder lines, Illus. 77, or from another radiator. Since the water is constantly being pumped through the system, then back to the boiler through a return line, all radiators maintain constant heat, except when an air block occurs.

A 2x3 or 2x4 SLEEPER

B REINFORCING WIRE

FEEDER

(77)

Many baseboard radiators have automatic air bleeders while the older ones, Illus. 78, require loosening valve A with a screwdriver. Air will hiss out of B. Since water will also spurt out, keep a cup or glass in position to catch it. When you remove air, and only water spurts out, the radiator is full.

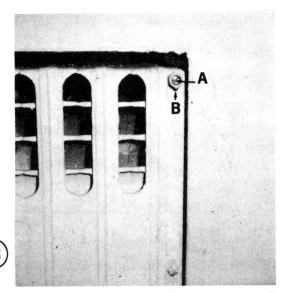

When a leak occurs in either the supply or return lines, or in any feeder, air will continually enter the system and you'll hear a gurgling or knocking.

When you have located the leak, switch burner off, close intake supply valve, open bleeder valves A, Illus. 78, on all radiators. Drain water from boiler. Drop pressure gauge to 0.

If you are purchasing a home that has hot water heating, make certain the seller supplies a map, or obtains one for you prior to purchase. Learn where and how deep hot water heating pipes are buried. Also find out where the main water and waste lines are buried. Some may be buried alongside, under or cross over heating lines.

Finding a buried pipe can be one of the most exasperating experiences a homeowner can go through, and anyone out to make a fast buck can legally rob you blind looking for a pipe.

Regardless of how much you value your time, no matter what you earn as a skilled craftsman or professional, the hourly rate charged for repairing a buried pipe will cost a sum far in excess to what you earn. If at all possible take a day or two of your vacation time and spend it on the job.

No matter what month you were born, and what luck you might have had in the past, whenever a buried pipe begins to leak, it's usually under your finest finished flooring or newest carpeting. If you are fortunate enough to have a map showing where each line is buried, you'll discover it's in a room with your most expensive wallpaper. Regardless of what you have previously spent on this room, don't lose your cool. Remember, you, and no one else caused the problem. Just believe that fate let you have one that happens to almost everyone at sometime.

Here's what to do. Remove all furniture, draperies, curtains, etc. Carefully remove the rug. Roll it, wrap it up and take it out of the room. Pry up the carpet strips. Make a drawing of the room so you can mark and number each piece of carpet strip. This simplifies replacing it in the exact same position.

Inspect floor to see if there are any water marks. If you find any, start tearing up the flooring. If you don't see any water damage do this.

As every homeowner who lives with a leaky roof soon learns, water flows in many mysterious and sneaky ways. When a pipe in a slab springs a leak, unless you are very rich, don't call in a plumber or heating contractor. Go first to your doctor and borrow a stethoscope. Ask him to show you how to use it, then probe your floor. Start probing where a hot water heating line enters the concrete. With the thermostat set high, and the circulator pumping water, probe the floor in the area where you think the pipe is buried. When you hear gurgling, you begin to zero in on the leaky spot.

As many floor slabs contain 2x3 or 2x4 sleepers, A, Illus. 77, and wire reinforcing B, you will need an electric hammer, a hand saw and wire cutters. Rent the electric hammer and wire cutters.

Use extreme care when using an electric hammer. Wear safety goggles and only use the hammer after the rental store has shown you how. An electric hammer is an easy tool for an

70

intelligent, alert person to use, but a dangerous tool for anyone who doesn't recognize its power.

Using a 1", or wider wood chisel, chisel a hole in one end of a piece of finished flooring. Make it large enough so you can drive a wrecking bar under the end, or side of an adjacent strip of flooring. You may have to chisel several holes in the same strip of flooring to get the first strip removed. Once you get started, you can save and reuse much of the flooring. When you have removed the finished flooring, and/or sub-flooring, start knocking a hole in the concrete some distance away from where you think the pipe is located. Study the feeder connections to radiators to make certain a feeder line isn't located in area selected. A feeder line is one that supplies each radiator. If possible, try to locate the heating or plumbing contractor who made the installation. He can save you considerable time, labor, mental anguish and money. By first tracing the line where it goes underground after leaving the boiler, you can frequently follow its exact direction by feeling the warmth or hearing the sound of water.

Proceed very cautiously until you break open a hole in concrete. Use care to break this hole clear of pipe. Use a hand hammer and chisel to chip hole larger until you can get your hand under slab. Dig earth out and keep feeling around until you find the pipe. Always use a hand hammer and chisel when you work near the pipe. Don't gamble using an electric hammer. One touch with an electric hammer and you need to replace more tubing than any leak normally requires.

When you find the leak, open up concrete so you or a plumber can cut out the leaking length of pipe and sweat two nipples and a length of tubing.

If you have to saw through a length of sleeper to make more work space, it's OK, and it doesn't need to be replaced.

Once you have located the pipe, make a chart showing distance from wall, its direction and depth below floor. Also indicate position of feeder lines.

Leave repaired pipe exposed until you run a two or three day test. When replacing fill, use only clean sand or dirt fill, and be sure it doesn't contain any nails, cinders or other foreign matters. Repatch concrete. Replace underlayment, etc., and say a prayer of thanks you could find and follow directions.

TELEPHONE SHOWER

One of the latest devices designed for better living is the "telephone" shower, Illus. 79. This easy to install, hand held shower permits washing any part without wetting others. It is particularly popular with women who want to shower without wetting their hair.

Those who don't have a bidet find this device almost as efficient in cleansing vital areas.

The telephone shower consists of a flexible hose that can be fastened directly to existing shower arm, Illus. 80. Or you can keep your shower and use the diverter connection, Illus. 81.

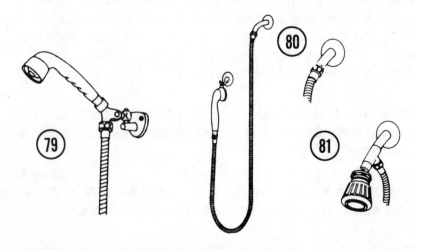

If you prefer to do away with showerhead and make a close fitting connection, a replacement elbow, Illus. 82, is also available. If your bath doesn't have a shower, remove discharge nozzle and replace with adapter spout, Illus. 83.

Be sure to insert washer between new fitting and threaded stud.

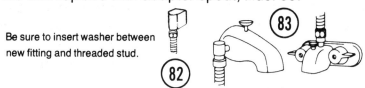

PLUMBING FIXTURE SIZES

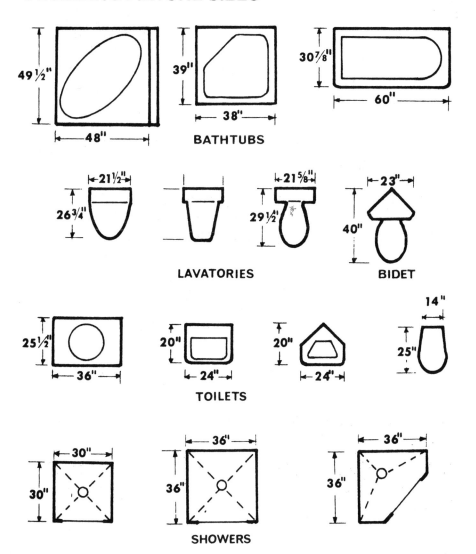

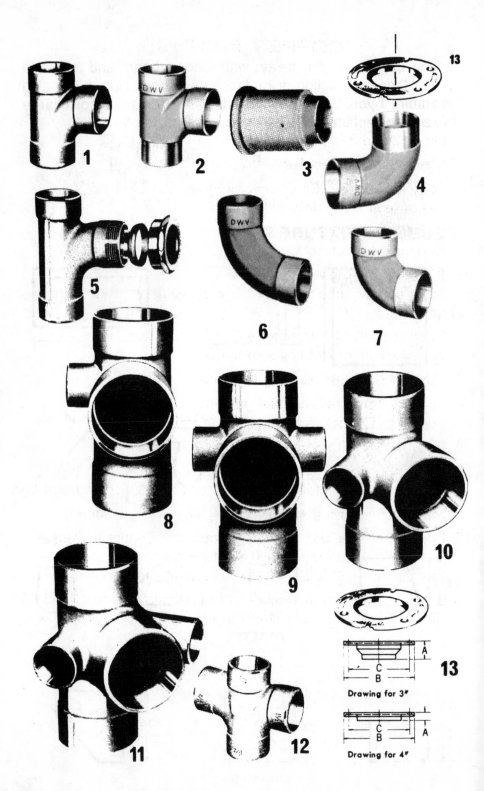

1

2

3

4

5

6

7

8

9

10

11

12

13

Drawing for 3"

Drawing for 4"

74

COPPER FITTINGS

1 Sanitary Tee. Copper to Copper to Copper.
Available in many different sizes. To indicate size specify number 1, 2 and 3 in order indicated.

2 Fitting Sanitary Tee. Fitting to Copper to Copper.

3 Soil Pipe Adapter. Connects Soil Pipe Hub to Copper tube.
Available in many different sizes.

4 Quarter Bend 90°. Copper to Fitting.
Also used as a closet bend with floor flange No. 13.

5 90° Sanitary Tee with Slip-Joint.
Joins copper tube to chrome or brass drain from lavatory or bathtub.

6 Quarter Bend 90°. Copper to Copper long turn.
Joins two lengths of copper tube.

7 Quarter Bend 90°. Copper to Copper.

8 Sanitary Tee with Side Inlet on Left.
Copper to Copper to Copper to Copper.

9 Sanitary Tee with Side Inlet on Left and Right.

10 Stack Fitting with Left Inlet.
Provides inlet for toilet, also 1½" or 2" inlet for drainage line.

11 Stack Fitting with Right and Left Inlet.

12 Double Sanitary Tee. Copper to Copper to Copper to Copper.
Used where code requires separate vents to main stack.
Available 3" x 3" x 1½" x 1½" and many other sizes.

13 Closet Flange attaches toilet to closet bend. Quarter Bend 90°, Illus. 4, can be used as a closet bend.

COPPER FITTINGS

14

Solder Cup End.

15

Fitting End. This end goes into fitting.

16

Usually designated FPT or FSPS. Female pipe thread or female standard pipe size.

17

45° Y-Branch. Copper to copper to copper. Joins main line with copper branch line.

18

45° Double Y-Branch.

19

Quarter Bend (90°) with side inlet. Copper to Fitting to Copper. Fitting End sweats to closet flange. Side inlet simplifies connecting drainage line from lavatory and/or bathtub.

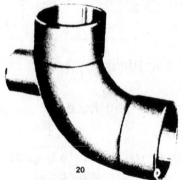

20

Quarter Bend (90°) with High Heel Inlet. Copper to copper to copper.

21

Fitting Reducer. Fitting to copper.

22

Coupling with Slip-Joint Connection. Joins copper tube to chrome or brass waste line from lavatory or bathtub.

23

Fitting Slip-Joint Adapter. Joins fitting to chrome or brass waste line from lavatory or bathtub.

COPPER FITTINGS

24
Male Adapter — Copper to MSPS.
Joins copper tube to threaded female.

25
Male Fitting Adapter — Fitting to MSPS.
Joins fitting to threaded female.

26
Coupling with stop. Copper to copper.

27
Repair coupling. No stop.

28
Roof Vent Increaser — Copper to Fitting.
3″ x 4″ x 18″ or 24″ or 30″ long.

29
Eighth Bend (45°) with cleanout,
used in line where cleanout is
required.

30
Vent Increaser — Copper to Copper.
Can be used to change diameter of vent stack just before going
through roof. Available in wide combination of sizes — 1½″ x
4″, 1½″ x 3″, etc.

32
Upturn MSPS fastens to bottom of one piece drum trap #33.
Bathtub outlet connects to Slip Joint.

31
Long Turn T-Y. Also available with side inlet on branch.

33
Drum Trap — One piece. Inlet from bathtub in bottom,
outlet in top.

34
90° Elbow — Copper to Copper.
Available in short or long radius.

36
45° Elbow — Copper to Copper.

38 Cap.

35
90° Ell Short Radius, also available long radius.
Copper to Fitting.

37
45° Fitting to copper ell.

77

CAST IRON

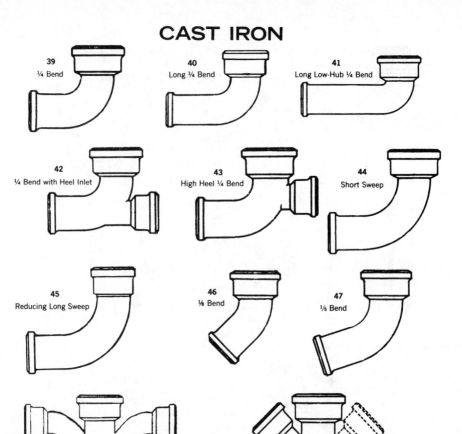

39
¼ Bend

40
Long ¼ Bend

41
Long Low-Hub ¼ Bend

42
¼ Bend with Heel Inlet

43
High Heel ¼ Bend

44
Short Sweep

45
Reducing Long Sweep

46
⅛ Bend

47
⅕ Bend

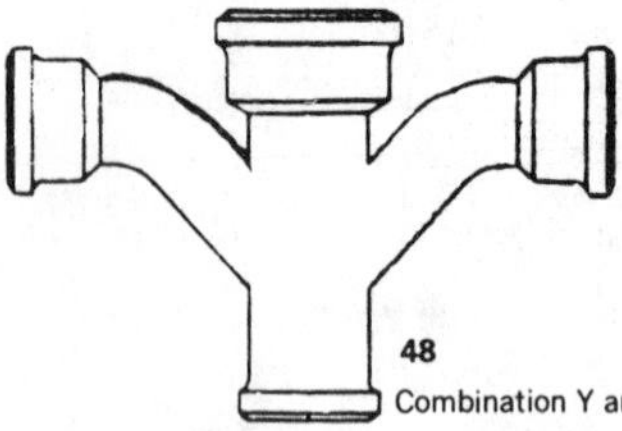

48
Combination Y and ⅛ Bend, Double

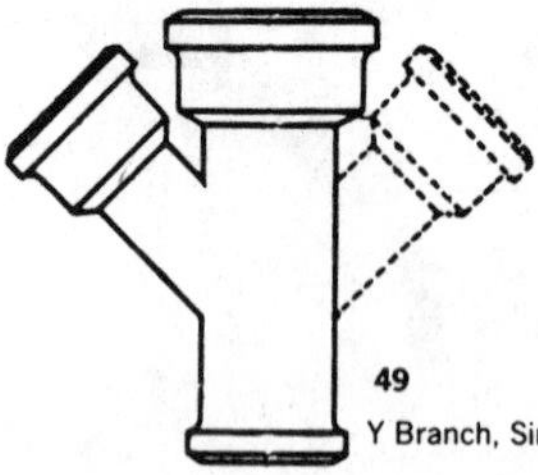

49
Y Branch, Single and Double

50
DIMENSIONS OF CAST IRON SOIL PIPE IN INCHES

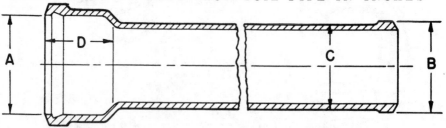

STANDARD				
SIZE	A	B	C	D
2	$2^{15}/16$	$2^5/8$	$2^1/4$	$2^1/2$
3	$3^{15}/16$	$3^5/8$	$3^1/4$	$2^3/4$
4	$4^{15}/16$	$4^5/8$	$4^1/4$	3

EXTRA HEAVY				
SIZE	A	B	C	D
2	$3^1/16$	$2^3/4$	$2^3/8$	$2^1/2$
3	$4^3/16$	$3^7/8$	$3^1/2$	$2^3/4$
4	$5^3/16$	$4^7/8$	$4^1/2$	3

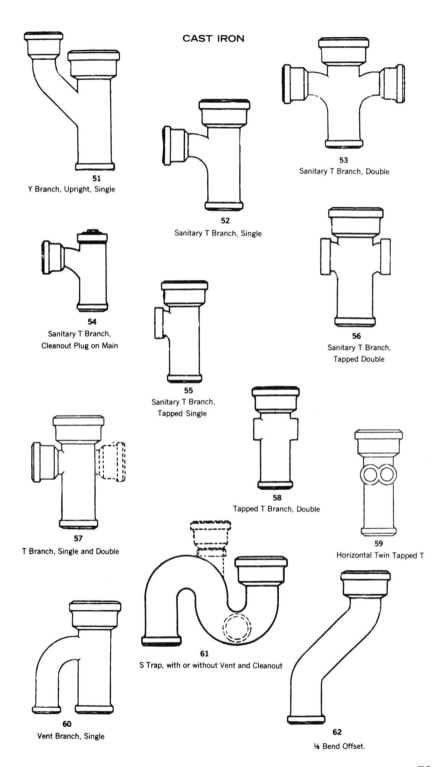

51
Y Branch, Upright, Single

52
Sanitary T Branch, Single

53
Sanitary T Branch, Double

54
Sanitary T Branch,
Cleanout Plug on Main

55
Sanitary T Branch,
Tapped Single

56
Sanitary T Branch,
Tapped Double

57
T Branch, Single and Double

58
Tapped T Branch, Double

59
Horizontal Twin Tapped T

60
Vent Branch, Single

61
S Trap, with or without Vent and Cleanout

62
⅛ Bend Offset.

GLOSSARY OF WORDS USED IN PLUMBING

FSPS   Female standard pipe size.

MPT   Male pipe thread.

MSPS   Male standard pipe size.

ADAPTER   A fitting used to connect two different size pipe or fitting. For Example: Soil Pipe Adapter joins cast iron soil pipe hub to 3", 2" or 1-1/2" copper tube. Available in many different sizes in both male and female.

BACKING BOARD   Usually a 1x4, 1x6 or 1x8 mortised in and nailed flush with leading edge of studs to provide support for a fixture.

BIDET   (pronounced be-day) A fixture that cleanses the essentials.

BRANCH   Any vent or drainage pipe other than soil or vent stack.

BRIDGING   A support or nailor nailed between joists or studs.

CALKING   Also spelled caulking. Material like oakum and lead.

CALKING TOOLS   iron, plumbers furnace, lead pot, ladle.

CAST IRON   Hub and Spigot.

CAST IRON CALKING RUNNER   Asbestos rope clamp. Permits pouring lead in horizontal hub and spigot joint.

CAST IRON PLUMBING TREE

CAST IRON PIPE CUTTER

CAT   Short piece of lumber, usually 2x4, nailed between studs to back up edge of gypsum board or fixture.

CHASE   Recess cut in framing to permit installing pipe.

CLEANOUT PLUG   cast iron.

CLOSET BEND

CLOSET FLANGE   Also called floor flange.

CLOSET SCREW   Long screw with detachable head formerly in wide use for fastening water closet to floor.

CORNER TOILET

COUPLING   A coupling joins two pieces of pipe of the same or different sizes. Some couplings have stops to allow pipe to only go in so far; others have no stop. Also available Copper to Slip-Joint.

DIVERTER VALVE   bathtub and shower control. Connects to hot and cold water line.

DRAINAGE   Any pipe that carries waste water in the drainage system.

DROP EAR ELBOW 90°   1/2" x 3/4". Permits connecting copper shower supply line to threaded nipple required for shower head.

DRUM TRAP   recommended for bathtub installation.

DWV FITTINGS   these drainage, waste and vent fittings incorporate the recommended drainage pitch of 1/4" to the foot.

ELL — L — ELBOW   a Quarter Bend 90°.

ESCUTCHEON   A plate used to enclose pipe or fitting at wall or floor opening.

FEMALE   end of fitting receives male.

FERRULE   a threaded sleeve soldered to hub of pipe.

FITTINGS   Any coupling, tee, elbow, union, etc., other than pipe. Plumbing catalogs refer to fittings as "ftg."

FIXTURE PLACEMENT CHART   p. 73.

FIXTURE UNIT   A method of estimating amount of water a fixture discharges. A unit is equivalent to 7-1/2 gallons of water or one cubic foot of water per minute. While a bathroom containing toilet, lavatory, bathtub or shower stall is rated by national codes as 6 units, the same codes rate a bathtub with 1-1/2" trap, with or without shower, 2 units; with 2" trap — 3 units; a bidet — 3 units; lavatory — 1 unit; shower stall — 2 units; an extra toilet with 3" drain — 4 units.

FPT   Indicates female pipe thread.

FRAMING   walls for bathroom.

FRESH AIR INLET   Pipe above roof. Codes frequently require this be size larger than internal vent line.

INCREASER   A coupling with one end larger than the other. Used to increase diameter of pipe above roof.

INSPECTION PANEL   provides access to bathtub trap.

KAYFER   Also called Kafir. A screw type hub fitting on cast iron that simplifies making new connection in 4" soil line.

LAVATORY P TRAP

LEAD BEND   formerly used exclusively as a closet connection to soil line.

MALE   End of fitting inserts in female.

NO-HUB CAST IRON

OFFSET   Any combination of pipe and fitting, or combination of fittings used to angle over.

PARTITION   end of tub.

PIPE CUTTER

PIPE MARKER   insures sawing pipe square.

PIPE SIZING CHART   Simplifies sizing existing pipes.

PIPE STRAP

PLUMBING TOOLS

PRE-ENGINEERED PLUMBING WALL   a completely assembled plumbing wall.

PRE-FABRICATED BATHROOM   see Book #682.

REDUCER   Copper to copper. Joins 3/4" to 1/2". Also available in other sizes.

ROOF VENT INCREASER   Fitting simplifies increasing vent from 3" to 4". Available 3" x 4" in 18", 24", 30" lengths.

ROUGH_IN   This describes installation of drainage waste and supply lines. Roughing-in concerns all work required prior to connecting fixtures.

SANITARY TEE 90° WITH SLIP JOINT   Joins copper waste and vent line to chrome or brass lavatory or sink drain pipe. Nut tightens lead ring to make tight joint.

SLIP JOINT

SOIL STACK   Codes allow 3" or 4" cast iron, copper, plastic.

SOIL LINE LAYOUT GUIDE   use folding rule or garden hose, see Book #682.

SOIL PIPE   carries discharge from one or more toilets and/or discharge from other fixtures to main sewer line.

SOLDER CUP END

SPIGOT   end of pipe that fits into hub.

STACK VENT OR VENT STACK   that part of the soil stack above the highest drain connected to stack.

STOP, VALVE   Available in angle and straight stops. Permits shutting off supply to fixture when repairs are required.

TOILET   roughing-in,.

TRAP   A fitting designed to provide a liquid seal to prevent back passage of air. P-Trap, Drum Trap, House Trap.

TUBE BENDER   tool simplifies bending soft copper tubing.

TUBE STRAP

VENT STACK   this is a vertical pipe that provides circulation of air to branch vents, revents or individual vents.

WALL HUNG TOILET

WASTE   this refers to water from any fixture except toilet.

WASTE PIPE   one that conveys only liquids, no fecal matter.

WET VENT   a wet vent is both a vent and drainage line from any fixture except a toilet.

Y-BRANCH

Write to Directions Simplified, Inc., P.O. Box 215, Briarcliff Manor, N. Y. 10510, for complete information concerning Easi-Bild Patterns and Directions Simplified Home Improvement Books In Canada, P.O. Box 4090, Postal Station A, Toronto, Ontario.

EASI-BILD 675/$1.50

PLUMBING
REPAIRS
SIMPLIFIED

DONALD R. BRANN

PQK124445

EASY-TO-FOLLOW
ILLUSTRATIONS

STEP-BY-STEP
DIRECTIONS